William B. Learned

The watchmakers' and machinists' hand book

William B. Learned

The watchmakers' and machinists' hand book

ISBN/EAN: 9783741166242

Manufactured in Europe, USA, Canada, Australia, Japa

Cover: Foto ©Andreas Hilbeck / pixelio.de

Manufactured and distributed by brebook publishing software
(www.brebook.com)

William B. Learned

The watchmakers' and machinists' hand book

THE

WATCHMAKERS' & MACHINISTS'

HAND BOOK,

OR

BEGINNERS' GUIDE.

Containing a Few Simple Rules and Explanations on the Relation
of Wheels to Pinions with Methods of
Figuring the Same.

BY

WILLIAM B. LEARNED,

Late Superintendent of the E. Howard & Co. Watch Factory,
Boston, Mass., U. S. A.

CHICAGO:
Geo. K. Hazlitt & Co., Publishers.

293872

PREFACE.

During the many years I have been employed in watch factory life, it has often been suggested by those I was endeavoring to instruct in the methods of figuring trains of wheels and pinions for watches, that if I would publish a book that could be purchased at a reasonable price, giving my methods and explanations in the same plain and simple manner in which they had been communicated to them, it would be highly appreciated by, and of great benefit to many who had been unable to obtain them in any plain and concise form from any other source.

At the present time, having a little leisure, I have decided to act upon these suggestions, as, looking back to the time when as an apprentice, I was most anxiously seeking information that would help me to understand how the different numbers of teeth in wheels and leaves in pinions were calculated, in order to obtain certain needed results, I should have been more than delighted.

If therefore I could do anything to assist those in like circumstances by helping them to obtain the information I had had such difficulty in procuring, I should at least be doing as I would gladly have been done by, and my labor would not be entirely in vain.

Could I at that time have obtained some book containing the simple rules and explanations by which I could have worked out some of the simpler problems, it would have been of very great advantage to me in more ways than one, for no one knows how soon, or at what period of his life he may be called upon to occupy a position in which this much needed information may be of great advantage to him.

My own experience was, that in all the works on horology that I was able to procure, (and I procured quite a number, at very great expense,) the problems were all purely algebraic, and the writers had given their whole attention to the more advanced portion of the science, leaving the beginner who may not have had the advantage of learning the higher branches of mathematics, to plod along as best as he could, until at last he became weary of trying to obtain the first rudiments and so decided to drop the matter entirely and do the best possible without this knowledge.

In offering this little volume to the public, the writer lays no claim to erudition or beauty of diction, neither does he claim to have any new or startling theories to offer. His sole and only desire is that he may be able to impart some little useful knowledge that may be of benefit to those desiring to learn, and thus save them some of the trouble and discouragement that beset him during the earlier period of his study of this most interesting subject.

The writer takes this method of acknowledging his great indebtedness to such writers as Reid, Arnold, Willis, Saunier, Frodsham and others, for much of the knowledge and information he may possess and hereby tenders to them his most sincere thanks.

W. B. LEARNED.

SIZING OF WHEELS AND PINIONS.

Upon the proper sizing of wheels and pinions, together with the correct form of teeth, largely depends the fine performance of the watch in which they are placed, as ever so finely finished a watch, with a poor train, can but be a constant source of trouble and annoyance to every one brought in connection with it, manufacturer, dealer, owner or repairer.

Charles Frodsham, in his "Elements of Watch and Clock Making," very wisely remarks that "the perfection of every watch or chronometer lies in the judgment and ability of the watchmaker to combine the several parts into one harmonious whole."

Let it not for a moment be understood that the *train* is the most important feature of a good watch, but with a *perfect* balance, balance spring and escapement, the best results cannot be obtained unless the power of the main spring can be properly transmitted through the train to these most important factors.

In another place in the same article Frodsham says: "The sizing of wheels and pinions is an operation of

0

daily requirement, and of the utmost importance in the production of good results." And again: "In paying attention to the *sizing* of wheels and pinions, we must give *equal* attention to the form of the teeth, or however correct the size, we shall not effect good gearing."

DEFINITIONS.

As this work is more particularly intended for beginners, or apprentices, rather than for finished workmen, it may be well to use a portion of our space in giving the proper definitions of some of the terms to be used, in order that they may have a full understanding of the meaning of them, as well as to have them convenient to be referred to as needed.

It is perhaps unnecessary for me to say that a

WHEEL—Is any circular piece of metal on the periphery of which teeth may be cut of various forms and numbers; while a

PINION—Is a smaller wheel, usually termed pinion, with teeth or leaves playing in connection with a larger wheel.

PITCH CIRCLE.—A circle concentric with the circumference of a toothed wheel, and cutting its teeth at such a distance from their points as to touch the corresponding circle of the pinion working with it, and have with that circle a common velocity, as in a rolling contact.

For example, imagine these two circles to represent the wheel and pinion, at their pitch circles, the outside, or dotted line, representing the points of the teeth or full diameter.

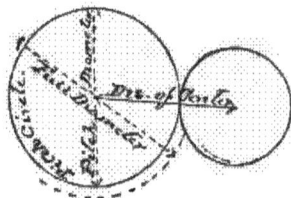

These circles should be so placed in relation to each other that when one is revolved—both being on their centers—the other should revolve with it by frictional contact.

PITCH DIAMETER.—The diameter of each of these circles is called the pitch diameter.

FULL DIAMETER.—The diameter measured from point to point of teeth is called the full diameter.

DISTANCE OF CENTERS.—The distance measured on a straight line from center to center, between the wheel and pinion, is called the distance of centers, and should be measured very accurately, as on this measurement the entire success of all calculations for a train of wheels depends.

LINE OF CENTERS.—A line drawn from center to center of any wheel and pinion, at which the two acting points should meet.

CIRCULAR PITCH.*—The pitch circle divided into as many spaces as there are teeth on wheel or pinion.

DIAMETRICAL PITCH.†—The *diameter* of the pitch circle, divided into as many spaces as there are teeth on wheel or pinion.

ADDENDA.—The portion of the tooth, either on wheel or pinion, outside of the pitch circle, is called the "addenda" or "working depth."

DRIVER.—Usually the wheel or the one that forces the other along.

DRIVEN.—Usually the pinion or the one that is being forced.

PROPORTION OF ADDENDA.

WHEEL AND PINION.

These proportions allude to the "time train." For the "dial train" other proportions are used. (See dial

---

*† You will observe that the circular pitch and the diametrical pitch are one and the same, *except* that one is the division of the *circle*, and the other the division of the diameter, each division indicating a tooth and a space.

train.) There seems to be some diversity of opinion, although very slight, among celebrated horologists in relation to the amount, as well as in the division of this addenda, as divided between driver and driven.

Arnold recommends two and one-quarter (2.25) to the driver, and one and one-half (1.50) to the driven, making a total of three and three-quarters (3.75), while Frodsham, although using the same amount (3.75) divides it a little differently, by giving two and one-half (2.50) to the driver and but one and one-quarter (1.25) to the driven. Many others claim that while Frodsham's rule is the better of the two for the driver, it is hardly sufficient for the driven, consequently they have adopted a total of *four* (4.00), dividing it thus: For the driver two and one-half (2.50), using Frodsham's rule for that, while they use Arnold's for the driven, viz., one and one-half (1.50). I have, however, given the preference to Frodsham, and adopted his rule, feeling that it is better to err in favor of having the driven a trifle small, rather than a little large.

To the dial train Frodsham gives to cannon pinion, as driver, two and one-quarter (2.25) "diametrical pitches," and to minute wheel, as driven, one and one-half (1.50), the same amount and proportion that Arnold gives to the time train, while to hour wheel he gives two and one-half (2.50), and to minute pinion one and one-half (1.50), making a total of four (4.00), the same amount and proportions mentioned as used by some for time train. In stem-winding watches, where the setting of the hands is operated through the setting train, and in that case the *drivers*, at times become the *driven*, it is thought best that the proportions should be equal, thus giving two (2.00) to each, driver and driven.

GEARING.

To those not familiar with gearing in general it may be well to state that the pitch circle of driver, together with the curve of the addenda outside said circle, does the work of forcing the driven along, and should be of such a form as to work easily and freely, with a continuous pressure until another tooth comes into operation on the line of centers. On the other hand the curve of the driven is " more for ornament than use," and as far as the action is concerned, might as well be cut off a little outside the pitch circle, and rounded over a trifle to prevent butting. The pitch circle of the tooth on driver, should not come into action with the driven until the line of centers is reached.   The thickness of the teeth in the driver should be such as to give them perfect freedom when between the leaves of the driven and pointing to center, yet not so much so as to give any lost motion, while it may be said of the driven that the leaves must not be so thick as to cause any bind in the space between the teeth of the driver when pointing to its center.   In fact there should be a very little shake in both.   Care must also be taken that the curve outside the pitch circle of the driver is not so long as that the point of the tooth shall strike the pinion at the root or bottom of the leaves.

EPICYCLOIDAL CURVE.

The epicycloidal curve is considered to be the most perfect working curve for the teeth of drivers, and there is no doubt that in all cases where there is plenty of room, it is all that is claimed for it, but the difficulty in using it for such small gearing as watch trains is, that the curve of the tooth on the driver is so long from the pitch circle to the point, that in such small pinions it is almost impossible to make the depth of the cut from the pitch circle to the root of the tooth, sufficiently deep to give the necessary clearance without cutting into the staff, or making the staff of the pinion so small as to be extremely weak. The nearer we can approach this curve however, and avoid the difficulties mentioned, the more perfect will be the train.

As before mentioned, it will not be necessary to follow this curve for the pinion. Any curve to suit the eye may be used, but great care should be taken to keep the pitch circle at its correct diameter, or the proper depthing may be destroyed.

PROPORTION OF WHEEL TO PINION.

The proportion of wheel to pinion must be, as the the number of teeth contained in wheel and pinion, and size of each, is to the distance of centers, within which they are to run.

The rule is universal, that whatever the number of teeth in wheel and leaves in pinion, and whether the ratio be whole or fractional, they must each have their proportionate part of the distance of centers. If a wheel is to have 60 teeth and a pinion 10 leaves, then the distance of centers must be divided between them, in this case, six to one, or six parts for the radius of the wheel, and one part for the radius of the pinion. For instance, sixty teeth divided by ten equals six, ten leaves divided by ten equals one, or one part of the distance of centers, using a common divisor for both. It follows therefore, that this distance of centers must be divided into *seven* equal parts or portions, giving to each its proper share. If the wheel is to have 75 teeth and the pinion 10 leaves, then the distance of centers must be divided into eight and one-half parts, giving to wheel seven and one-half parts, and to pinion one part.

FIGURING TRAINS.

Having given these necessary definitions and explanations so that the student may understand what may follow, we will proceed to explain in the same simple manner the *modus operandi* of figuring a train of wheels and pinions for a watch beating 18,000 times an hour, running thirty-eight hours, with once winding, on five turns of the stop work.

The question may be asked, how is the number of teeth in wheels and leaves in pinions to be determined, or decided upon and why should the fourth wheel contain more teeth than the third wheel, which is the larger

of the two? To many this may seem a very foolish
question but they must remember that *they* at some
period of their apprenticeship probably asked *many* just
such foolish questions and also remembering that this
work is *particularly* intended for beginners, I trust
they will pardon the intrusion, and if they do not care to
read it, they can pass it over as lightly as they please.
The answer is this: In all watches denoting seconds, the
fourth pinion must be made to revolve sixty times to one
revolution of the center pinion, which revolves once
every hour.   Such numbers of teeth must therefore be
used as will accomplish this result, and any numbers that
will accomplish it will be found to be correct.

Before we can show the manner of procedure we
must decide upon a train of wheels and pinions that we
wish to obtain the dimensions of.   Suppose for this purpose we take the following:

Main wheel of 76 teeth, center pinion 10 leaves.

Center wheel of 80 teeth, third pinion 10 leaves,

Third wheel of 75 teeth, fourth pinion 10 leaves.

Fourth wheel of 80 teeth, escape pinion 8 leaves.

The center distances we will imagine to be as follows,
viz.:

Main to center, .450; center to third, .323; third to fourth, .272; fourth to escape, .243. As but one-half of the diameter of each wheel and pinion is contained within the given distance of centers and we are to determine the pitch diameter of each, it becomes necessary for us to first multiply this distance by two and divide the quotient by the sum of the teeth in wheel and leaves in pinion which will give us the "diametrical pitch." We will proceed first with the main and center. Example 1, (with explanations),

Main wheel,   76 teeth, ⎫
Center pinion, 10 leaves, ⎬ Distance of center, .450.
                          ⎭
         ___
         86

          .450
            2
       86).900(.010465—Diametrical pitch.
          86
         ___
         400
         344
         ___
         560
         516
         ___
          440
          430
         ___
           10

Having obtained this " diametrical pitch," we will proceed to obtain the necessary " addenda " for the wheel.
This you will remember is to be two and one-half (2.5)
diametrical pitches, therefore, diametrical pitch, multiplied two and one-half times equals addenda of wheel.

$$.010465$$
$$\underline{2.5}$$
$$052325$$
$$\underline{020930}$$
.0261625 Addenda for wheel.

The "pitch diameter" of the wheel must be as many
"diametrical pitches" as there are teeth on the wheel,
therefore, the diametrical pitch multiplied by number of
teeth, as

$$.010465$$
$$\underline{76}$$
$$062790$$
$$\underline{73255}$$
.795340= Pitch diameter.

giving us, therefore, a main wheel with a "pitch diameter" of .795340.

Now to obtain the "full diameter" we must add the
" pitch diameter" and the " addenda " together, being

very careful to place the decimals one under the other correctly, thus,

Pitch diameter of wheel,	.795340
Addenda,	.0261625
Full diameter of main wheel,	.8215025

We will now proceed in the same manner to obtain these two diameters for the pinion, which must have *ten* "diametrical pitches" for its "pitch diameter," with the proper "addenda" added, for its "full diameter."

Diametrical pitch .010465 multiplied by
number of leaves   10 results in
_____
                    .104650

giving us a pitch diameter for pinion of .104650. Now proceed to obtain the "addenda" for pinion. This you will remember is to be one and one-quarter (1.25) "diametrical pitches," therefore

$$\begin{array}{r} .010465 \\ 1.25 \\ \hline 052325 \\ 020930 \\ 010465 \\ \hline .01308125 = \text{Addenda to pinion.} \end{array}$$

This added to pitch diameter .10465
                        Addenda .01308125
                                _____
gives us a full diameter   .11773125

Now let us try and prove our work and see if we have done it correctly. If we have, by adding the two pitch diameters together, and dividing the quotient by two, we shall have as a product the "distance of centers."

Pitch diameter of wheel, .795340
Pitch diameter of pinion, .104650

$$2).899990$$

Distance of centers,   .449995

You will notice that I have carried these products out to the extent of six, or even in some cases, to eight decimals, and yet have failed to obtain the *exact* figures. This comes from the fact that we did not carry our figuring out far enough to obtain a quotient *without a remainder*. Had we done so, the result would have come out just right, viz.: .450.

In figuring for a train, the nearer you can come to a product without a remainder, the closer you can prove your work, but in actual measurements, such as in diameters, center distances, etc., you will only be expected to obtain the nearest thousandth (three decimals.) For instance, take the two pitch diameters just obtained, first of the wheel .795.340. As the fourth figure in the decimal is less than 5 or one half, you may with safety call it .795, while, with the pinion, .104,650, the fourth figure being more than 5, I should advise giving the benefit to the preceding figure, calling it .105, for the pitch diameter of the pinion.

Having given the first example, and explanations in detail, so that the beginner may receive the explanation with the figures, I will proceed to give the same example "en masse:"

EXAMPLE I.

(Without explanations.)

Main wheel, 76 teeth,
Center pinion, 10 leaves, } Distance of centers, .450.

```
                86
          .450—Distance of centers.
                 2
Sum of teeth, 86) 900 (.010465—Diametrical pitch.
                 86
```

```
                400        .010465
                344             76 Teeth in main wheel.
.010465         560        062790
    2.5         516        073255

052325          440        .795340  Pitch dia. of main wheel.
020930          430       .0261625  Addenda "    "       "

.0261625 Addenda 10        .8215025  Full dia. of main wheel.
    for wheel.
```

```
.010465      .795340          .010465
  1.25       .104650           10 Teeth in center pinion.
 ─────       ───────         ──────
052325       2).899990        .104650 Pitch dia.   "      "
020930        ───────         .01308125 Addenda "       "
010465        .449995 Dis.    ─────────
─────────           of cen.   .11773125 Full dia. "       "
.01308125 Addenda for pinion.
```

The result shows as follows:

Pitch dia. main wheel,   .795340   Full dia.   .8215025
"   " center pinion,   .104650   "   "   .11773125

Next in course come Center Wheel and Third Pinion, which require the same method of procedure.

Center wheel, 80 teeth, } Distance of centers, .323
Third pinions, 10 leaves, }
      ───
      90 == Sum of teeth.

.323 Distance of centers.
   2
  ───
90) .646 (.007177 + = Diametrical pitch.
    630
   ────
    160
     90
   ────
    700
    630
   ────
    700
    630
   ────
     70

.007177 = Diametrical pitch.

2.5
___

035885
014354
___

.0179425 = Addenda of wheel.

.007177
80   Number of teeth in wheel.
___

.574160   Pitch diameter of wheel.
.0179425 Addenda   "   "
___

.5921025 Full diameter   "   "
.007177 = Diametrical pitch.

1.25
___

035885
014354
007177·
___

.00897125 = Addenda of pinion.
.007177 = Diametrical pitch.
10   Number of leaves in pinion.
___

.071770   Pitch diameter of pinion.
.00897125 Addenda   "   "
___

.08074125 Full diameter   "   "

PROOF.

.574160   Pitch diameter of wheel.
.071770   "   "   " pinion.
___

2) .645930
___

.322965 = Distance of centers.

The result shows as follows:

Pitch dia. of center wheel, .57416   Full dia., .5921025
"   "   " third pinion,   .07177   "   "   .08074125

Following in course come the Third Wheel and Fourth Pinion.

Third wheel,   75 teeth, } Distance of centers, .272
Fourth pinion, 10 leaves, }

                 85

.272
2
Sum of teeth, 85).544 (.0064 Diametrical pitch.
                510
                340
                340

.0064   .0064
2.5   75
0320   0320
0128   0448
.01600 Addenda for   .4800 Pitch dia. of third wheel.
          wheel.   .0160 Addenda.
                        .4960 Full dia. of third wheel.

	PROOF.	
		.0064
		10
.0064	.4800	.0640 Pitch dia. of fourth pinion.
1.25	.0640	.0080 Addenda.
0320	2).5440	.0720 Full dia. of fourth pinion.
0128	272 Dis. of cen.	
0064		
.008000 Addenda for pinion.		

This result shows as follows:

Pitch dia. of third wheel, .4800; full dia., .4960.
Pitch dia. of fourth pinion, .0640; full dia., .0720.

Now comes the last of the Time Train, viz: the Fourth Wheel and Escape Pinion.

Fourth wheel, 80 teeth, } Distance of centers, .243
Escape pinion, 8 leaves, }
88

.243
2

Sum of teeth, 88) .486 (.005522—Diametrical pitch.

	440	
.005522	460	
2.5	440	
027610	200	.005522
011044	176	80
.0138050 Addenda	240	.441760 Pitch dia. of fourth wheel.
for wheel.	176	.013805 Addenda.
	64	.455565 Full dia. of fourth wheel.

28   THE WATCHMAKERS' AND

```
                              .005522
              PROOF.             8
.005522      .441760       .044176  Pitch dia. of escape pinion.
  1.25       .044176       .0069025 Addenda.
027610      2)485936       .0510785 Full dia. of escape pinion.
011044       .242968 Dis. of centers.
005522
.00690250 Addenda for pinion.
```

This result shows, as follows:

Pitch dia. for fouth wheel,   .44176; full dia., .455565.
Pitch dia. for escape pinion, .044176; full dia., .0510785.

Before taking up the dial train, and while the escape pinion is fresh in our mind, it may be well to spend a little time in a few

REMARKS ON AN EIGHT LEAF PINION.

It is conceded by the best horologists that an eight leaved pinion, especially as *small* a pinion as is required in a watch train, is a very difficult one to so cut that it will run perfectly free with the fourth wheel, having the usual addenda, as the point of the incoming tooth of the wheel is very liable to touch, or lodge, on the pitch line of the next approaching leaf of the pinion, particularly so, if the leaves of the pinion are a *very trifle* thick.

The writer has found, in his own experience, several cases where the watch would stop, occasionally, without any apparent provocation, and on the slightest jar start to running again, leaving it quite a little behind time, yet when noticed would be running as usual. In most cases of this kind, it has occurred while the watch was standing on the rack, or hanging on the watch board. After much trouble and annoyance in hunting for the cause, it was found to be, not from an over sized wheel or pinion, but from the very cause mentioned, viz: the point of the incoming tooth of the fourth wheel was caught on the pitch line of the incoming leaf of the escape pinion, causing the tooth and leaf to "*butt*" together just enough to *occasionally* stop the watch. This *might* have occurred from the cutter being a little thin in the first place, but in that case, *all* the pinions cut, with that cutter, would have had the same trouble, and all watches having that lot of pinions in them would have been similarly affected. As it was only *occasionally* that this trouble was found, it is but fair to suppose that it came from the wearing of the cutter, which had been used a little too long, and should have been taken for a *first* cutter or thrown away. It may seem strange to those not acquainted

with the manufacture of watch pinions, when they are
informed, that however good a cutter you may have,
the number of pinions you can cut with it, and have
them perfect, is quite limited. The sides of the cutter,
as well as the point, are constantly wearing, and, of
course, as they wear, the space between the leaves is
constantly growing narrower, and the leaf thicker.
The point can be sharpened by grinding back the face,
but the sides can never be touched after the cutter is
first made and hardened, consequently, great care
should be taken to see that it is not used too long, and
the leaf allowed to grow too fat.

Knowing, then, the natural difficulty with an eight
leaved escape pinion, together with this constant wear of
the cutter, the writer, having tried it thoroughly, and
adopted it himself, would recommend the lessening of
the addenda of the fourth wheel a very little, being, of
course, *very* careful to keep the pitch circle at its proper
size.

In trying this experiment, I took *two* "diametrical
pitches" for the fourth wheel, instead of two and one-
half, and in order to show the *very* slight difference it
makes in the full diameter of the wheel, I will proceed

to figure the fourth wheel and escape pinion, same as before, with the exception of the addenda.

Fourth wheel, 80 teeth, $\Big\}$ Distance of centers, .243
Escape pinion, 8 leaves,

88

.243
2

Sum of teeth, 88).486 (.005522—Diametrical pitch.
440

460
440

200   .005522
176   80

.005522   240   .441760 Pitch dia. fourth wheel.
2   176   .011044 Addenda.

.011044 Addenda   64   .452804 Full dia. fourth wheel.
for wheel.
.005522
8

PROOF.   .044176 Pitch dia. escape pinion.
.005522   .441760   .0069025 Addenda.
1.25   .044176
.0510785 Full dia. escape pinion.
027610   2).485936
011044
005522   .242968 Dist. of cen.

.00690250 Addenda for pinion.

It will be observed on comparison, that every set of figures, and their results, are precisely the same, except the addenda and the full diameter of the wheel, which is only about one one-thousandth smaller on the side in action, and yet it was all that is required, unless the leaf is out of all proportion to the space.

In my own experience, after making this change, I had no further trouble, neither could I observe any difference in the performance of the watches, except that they never stopped afterwards, or gave any trouble whatever. While on this subject of the escape pinion, let me speak of another matter connected therewith, and relate a little experience of my own, which *may* be of benefit to repairers, and save them more or less trouble. This relates more particularly to the care which should be exercised in the examination of watches, before repairing them. Let me therefore caution the repairer, especially if he be a young workman, to be *very* sure when he examines a watch for repairs, to look at every part very carefully to see that there are no worn parts that would be likely to cause trouble when the watch is cleaned, and put up ready for use. Among other things, be very careful to examine the escape pinion just where

the tooth of the wheel impinges the side of the leaf, to
see if the leaf is *worn* at that point, for if it is, he might
just as well make up his mind that he must put in a new
pinion, if he wishes to avoid much trouble and annoyance, for in nine cases out of ten, if he leaves it in he
will never be able to get the watch to run at all satisfactorily, and his customer will be constantly complaining.

The writer distinctly remembers one case in particular,
in my early experience as a repairer, when a very *crusty*
old gentleman, who owned a very fine Frodsham watch,
came into the store where I was at work, and handing
me the watch, which was then running, asked me what
was the matter with it, at the same time making the
remark that "he did not believe these *damned watchmakers* knew their business, anyhow, for he had had
that watch in the hands of several of the best watchmakers in the city, and none of them had succeeded in
repairing it, so but that it would stop every once in a
while, and then get to running again in his pocket without his knowing when it stopped, or that it was stopped
at all, except that when he took it out of his pocket to
look at the time he would find it some minutes behind
time, and it would be *sure* to occur just before he was

expecting to take a train, and the consequences would
be he would get left." When he informed me of the
firms who had had the watch, and knowing that they
employed some very fine workmen, I felt very much
disinclined to tackle it, but finally told him I would take
it down and examine it, and if I thought I could repair
it so it would perform satisfactorily, I would do so, and
he need not pay for it until he was entirely satisfied with
its performance. This seemed to please him very much,
and he left it with me. Suffice it to say, I repaired it by
putting in a new escape pinion, (which I found worn in
the leaves at the point mentioned,) cleaning, etc., and
never after did it give him any further trouble. The
result was, he not only paid me well for it, but he sent
me many customers by relating the incident to his
friends, and inducing them to bring their watches to me
for repairs.

I do not mention this incident to give the impression
that I was any *better* watchmaker, if as good, as those
who had repaired it before, but only with the idea of
impressing upon the mind of the beginner the necessity
of being *thorough* and conscientious in their work.
First, find out what there is to *cause* trouble, and then

remove the cause. Trusting that you will pardon this
digression, we will now resume our subject and proceed,
taking up next the

DIAL TRAINS.

This consists of the cannon pinion, minute wheel, hour
wheel and minute pinion.

For this train, Frodsham, as before stated, gives to
cannon pinion—as driver—an addenda of two and one-
quarter (2.25) "diametrical pitches," and to minute
wheel—as driven—one and one-half (1.50), the same
amount and proportion that Arnold gives to time train,
while to hour wheel he gives two and one-half (2.50)
and to minute pinion one and one-half (1.50), making a
total of four (4), the same amount and proportion mentioned as being used by some for time train.

In stem setting watches, when the setting of the hands
is operated through the setting train, and in that case,
the *driver* at times becomes the *driven*, it is thought
advisable that the proportions be made equal, thus giving two (2) to each *driver* and *driven*. As nearly, if not
not all watches made at the present time, are stem winding, I shall in these calculations use the above addenda.

In figuring for dial trains, the wheels and pinions may
be of any number of teeth and leaves, that when the
number of teeth in wheel is divided by the number of
leaves in pinion, that runs in connection with it and the
two quotients are multiplied together, the result shall be
twelve (12), for instance:

Cannon pinion of ten (10) leaves.
Minute wheel of thirty (30) teeth.
Minute pinion of eight (8) leaves.
Hour wheel of thirty-two (32) teeth.

The minute wheel of 30, divided by the cannon pinion
of 10 equals 3.

The hour wheel of 32, divided by the minute pinion
of 8, equals 4.

These two products multiplied together equal 12.

$$
\left.
\begin{array}{l}
\text{If for cannon pinion of } 12 \atop \text{Minute wheel of } 48 \Big\} = 4 \\
\text{Minute pinion of } 10 \atop \text{Hour wheel of } 30 \Big\} = 3
\end{array}
\right\} 12
$$

$$
\left.
\begin{array}{l}
\text{If for cannon pinion of } 14 \atop \text{Minute wheel of } 42 \Big\} = 3 \\
\text{Minute pinion of } 12 \atop \text{Hour wheel of } 48 \Big\} = 4
\end{array}
\right\} 12
$$

$$
\left.
\begin{array}{l}
\text{If for cannon pinion of } 16 \atop \text{Minute wheel of } 48 \Big\} = 3 \\
\text{Minute pinion of } 12 \atop \text{Hour wheel of } 48 \Big\} = 4
\end{array}
\right\} 12
$$

Having decided which you will adopt proceed in figuring, same as for time train, except in the amount of addenda to be given. We will, as an example, take the first dial train mentioned, viz:

Cannon pinion of ten leaves, and minute wheel of thirty teeth.

Minute pinion of eight leaves and hour wheel of thirty-two teeth.

The distance of centers we will call .163. Any other distance could just as well be used, but this will answer for illustration.

First:

Cannon pinion, 10 leaves, } Distance of centers, .163.
Minute wheel, 30 teeth, }

                    40
    .00815   .163
       2   2
    ──────   ──────
    .01630 Addenda 40).326(.00.815 Diametrical pitch.
       for both.   320
                   ────
                     60   .00815
                     40   10
                   ────   ──────
                    200   .08150 Pitch dia. for can. pinion.
                    200   .01630 Addenda.
                            ──────
                            .09780 Full dia. for can. pinion.

PROOF   .00815
.08150   30
.24450   ─────
─────   .24450 Pitch dia. for min. wheel.
2).32600   .01630 Addenda.
─────   ─────
.163 Distance of centers.   .26080 Full dia. for min. wheel.

Second.

Minute pinion, 8 leaves, } Distance of centers, .163.
Hour wheel,   32 teeth, }

                40
               .163
                 2
          ─────
    40).326(.00815 Diametrical pitch.
         320
         ───
          60
          40
          ───
.00815   200   .00815
   2   200   8
─────   ─────
.01630 Addenda for both.   .06520 Pitch dia. min. pinion.
                             .01630 Addenda.
                             ─────
                             .08150 Full dia. min. pinion.
PROOF.   .00815
.06520   32
.26080   ─────
─────   .01630
2).3260   .2445
─────   ─────
.163 Distance of centers.   .26080 Pitch dia. hour wheel.
                             .01630 Addenda.
                             ─────
                             .27710 Full dia. hour wheel.

Should you wish to adopt any other numbers in dial wheels and pinions proceed as above, and follow the instructions under the head of dial train, as the same rule must be followed in either case.

PROBLEMS.

We will now proceed to work out a few problems, which will show the proper manner of finding certain results, those results to be determined according to what is given and what is required.

PROBLEM I.

Given.—The number of teeth in wheel and pinion, with full diameter of wheel and distance of centers.

Required.—The pitch diameter of wheel, also pitch diameter and full diameter of pinion.

We will take, as an example, the same main wheel and center pinion as used in our first example in figuring for train wheels and pinions, viz., main wheel, 76 teeth, with full diameter of .8215; center pinion of 10 leaves, and distances of centers of .450. Proceed, as in figuring for regular train, by adding the number of teeth in wheel and leaves in pinion, and obtain the "diametrical

pitch" and the "addenda." Then take the given full
diameter of the wheel, .8215, and subtract the addenda
from it, which will give the "pitch diameter" of wheel.
Making use of the rule of proportion, proceed to obtain
the pitch diameter of the pinion, thus:

As the number of teeth in wheel is to its pitch diameter, so is the number of leaves in pinion to *its* pitch
diameter. Then add the proper addenda, and the result
will be the full diameter of the pinion.

EXAMPLE.

Main wheel, 76 teeth, } Distance of centers, .450.
Center pinion, 10 leaves,

```
            86                      .450
                                      2
  .010465              86).900(.010465 Diametrical pitch.
    2.5                 86
  _____               ____
  052325                400
  020930                344
  _____               ___
  .0261625 Addenda for wheel.  560
                               516
  .010465                      ___
    1.25                       440
  _____                       430
  052325                       ___
  020930                        10
  010465
  _____               .8215 Full dia. of wheel.
  .01308125 Addenda for pinion.  .0261625 Addenda.
                        _____
                        .7953375 Pitch dia. of wheel.
```

76 : 7953375 :: 10
10

76)7.9533750(.1046496 Pitch dia. of pinion.

Proof. 76 .01308125 Addenda.

.7953375 353 .11773085 Full dia. of pinion.
.1046496 304

2)8999871 493
.44999355 Dis. of cen. 456

377
304

735
684

510
456

54

This result shows that we have obtained what was "required," viz., the pitch diameter of the wheel, .7953375, together with the pitch diameter, and full diameter of the pinion; also, have proved our work by obtaining the same distance of centers with which we started. I have carried out the decimals much further than usual in order to show how close a result can be obtained.

PROBLEM II.

Given.—The number of teeth in wheel and pinion and full diameter of pinion—.11773—with distance of centers.

Required.—The pitch diameter of pinion, together with pitch diameter and full diameter of wheel.

Main wheel,   76 teeth, } Distance of centers, .450
Center pinion, 10 leaves, }

```
        86            .450
                        2

  .010465        86).900(.010465 Diametrical pitch.
    2.5           86

  052325          400
  020930          344          .11773 Given full dia. of pinion.
 .0261625 Addenda for wheel. 560          .01308 Addenda.
                  516          .10465 Pitch dia. of pinion.

  .010465         440
    1.25          430
  052325           10
  020930
  010465                    10 : 10465 :: 76
 .01308125 Addenda for pinion.        76

                                    62790
                                    73255

    PROOF.                      10)795340

   .10465 ·                       .79534 Pitch dia. of wheel.
   .79534                         .02616 Addenda.

 2).89999                         .82150 Full dia. of wheel.

   .449995 Distance of centers.
```

PROBLEM III.

Given.—The number of teeth in wheel and pinion, with distance of centers.

Required.—The pitch diameter and full diameter of both wheel and pinion.

Main wheel,   76 teeth, } Distance of centers, .450.
Center pinion, 10 leaves, }

```
        86              .450
                          2
 .010465 Dia. pitch.   86).900(.010465 Diametrical pitch.
      2.5              86
 ─────                 ──
 .052325               400
 020930                344
 ─────────             ───
 .0261625 Ad'nda for wheel.  560
                        516
                        ───
   .010465 Dia. pitch.  440     .010465 Dia. pitch.
        1.25            430          76
   ─────────            ──     ─────────
   052325               10     062790
   020930                      073255
   010465                      ─────────
   ─────────                   .795340  Pitch dia. of wheel.
   .01308125 Addenda for pinion.  0261625 Addenda  "    "
                                  ─────────
                                  .8215025 Full dia.  "    "

        PROOF.                   .010465 Dia. pitch.
     .795340                          10
     .104650                     ─────────
     ────────                    .104650 Pitch dia. of pinion.
     2).89990                    .01308125 Addenda.
     ────────                    ─────────
     .449995 Dis. of centers.    .11773125 Full dia. of   "
```

PROBLEM IV.

Given.—The number of teeth in wheel and leaves in pinion with pitch diameter of both wheel and pinion.

Required.—The distance of centers.

Main wheel, 76 teeth, pitch diameter, .79534

Center pinion, 10 leaves, "   "   .10465

Add the pitch diameters together and divide by two.

     .79534 Pitch diameter of wheel.
     .10465   "   "   " pinion.
     ‾‾‾‾‾‾‾
     2).89999
     ‾‾‾‾‾‾‾
     .449995 Distance of centers.

This result shows the distance of centers at which the gears will run correctly.

LOST WHEELS AND PINIONS.

We will now consider the problem of how to find the correct diameters of a wheel and pinion that is lost, together with the number of teeth and leaves, having the connecting wheels and pinions as a guide. Take for instance, a watch with the third wheel and pinion missing. First, count the number of teeth in center wheel and leaves in fourth pinion. Suppose we find the number of teeth in center wheel to be eighty (80) and the

number of leaves in fourth pinion to be (10). As every
fourth pinion in a watch denoting seconds must make
sixty (60) revolutions to one (1) of the center, the number of teeth in wheel and leaves in pinion must be so
calculated as to produce that result. If the center wheel
has eighty (80) teeth, then the third pinion must have
such number of leaves as will divide the number of
teeth in center wheel without a remainder. In this case
it can be done by using either (10) or eight (8) as a
divisor. If we use ten (10) the product will be eight
(8), showing *that* to be the number of leaves in the
third pinion, but as the fourth pinion has ten (10) leaves,
and on counting the center pinion we find that also to
contain the same number, it is hardly reasonable to suppose that an eight (8) leaved pinion would be put in
between two of ten (10) leaves. If we use eight (8) as
a divisor we obtain a product of ten (10) leaves for third
pinion, which is much more reasonable to suppose was
the original number, as it corresponds with the other
pinions found in the watch. We must, therefore, come
to the conclusion that a third pinion of ten (10) leaves
is what is needed to take the place of the one that is lost.
This being decided, it is *also* decided that the third pinion

will revolve eight (8) times to one revolution of the center, which gives one of the factors in the problem. Now, we must have such number of teeth in third wheel as to cause the fourth pinion to revolve as many times as is necessary to produce the required result, viz., sixty (60) revolutions of fourth pinion to one of the center.

Having obtained *one* factor (8), the other is obtained by dividing the number of revolutions the fourth pinion must make (60) by the factor already found, viz. (8) thus—60÷8=7.5. If therefore we multiply the number of leaves in fourth pinion (10) by the last found factor (7.5) we shall obtain the necessary number of teeth for third wheel, thus—10×7.5=75, as the number of teeth for the missing wheel.

We will now proceed to measure the distance of centers, and determine the diameters of both wheel and pinion.

In order to measure the distance of centers correctly, I should advise that a pair of parallel dividers be used, as will be explained under the head of "measuring tools" further on. Great care should be used in measuring these distances, to get them as exact as possible, for upon these depend the entire success of our calculations.

Having completed these measurements and found them to be as between the center and third .323—and between the third and fourth .272—proceed as in figuring the regular train. We know by counting that the center wheel has eighty (80) teeth, and have determined that the third pinion must have ten (10) leaves, with a distance of centers between them of .323. First proceed to find the "diametrical pitch," and then "pitch diameter" and "full diameter" of the missing pinion.

Center wheel, 80 teeth, } Distance of centers, .323.
Third pinion, 10 leaves, }

```
                    90

                          .323
                             2
                            ───
       .007177            90).646(.007177 Diametrical pitch.
         1.25               630
        ──────              ───
        035885              160
        014354               90
        007177              ───
      ──────────            700
      .00897125 Addenda for pinion.  630
                            ───
                            700       .007177
                            630         10
                            ───       ────────
       .007177  .           70        .071770 Pitch dia. of  pinion.
         2.5                          .00897125 Addenda.
       ──────                         ──────────
       .035885                        .08074125 Full dia. of     "
       .014354
      ─────────
      .0170425 Addenda for wheel.
```

.007177
80

.071770   .574160 Pitch dia. of wheel
.574160   .0179425 Addenda.

2).645930   .5921025 Full dia. of   "

.322965 Distance of centers.

The third pinion therefore must be of ten (10) leaves having a pitch diameter of .07177 and a full diameter of .08074125.

The question may be asked. How do you know that the center wheel you find in the watch has just this amount of addenda which you have figured? In answer I do not, as there is no *possible* way to measure it correctly. I *only* know that it *should* have it if made properly in the first place, and consider this the most accurate manner of obtaining it.

We will now try and decide what the "pitch diameter" and "full diameter" of the missing wheel should be. We have counted the leaves in fourth pinion, and found it to contain ten (10). We have also determined that the third wheel should contain seventy-five (75) teeth, and having measured the center distance, we find it to be .272—therefore proceed as before.

Fourth pinion, 10 leaves, ⎫ Distance of centers, .272.
Third wheel,   75 teeth, ⎬
                   ___
                    85

.0064   .272
 2.5   2
_____   _____
.0320   85).544(.0064 Diametrical pitch.
.0128   .510
_____   _____
.01600 Ad'nda for wheel.   .340   .0064
                               .340   75
                               _____   _____
                                          .0320
                                          .448
                                          _____
                                          .4800 Pitch Dia. of third wheel.
                                          .0160 Addenda.
                                          _____
                                          .4960 Full Dia. of third wheel.

As a full product then, of this calculation, we find that
we must supply a third wheel of seventy-five (75) teeth,
having a " pitch diameter " of .480, and "full diameter"
of .496, together with a third pinion of ten (10) leaves,
having a "pitch diameter " of .072, and " full diameter "
of .081.

MEASURING TOOLS.

The tool most commonly used for obtaining the distance of centers, is the ordinary depthing tool, as that

is an instrument that nearly every watchmaker has at
hand. It has however many objectionable features,
which to my mind, are very serious defects.

Out of the many in use, but very few can be found,
that the holes for the arbors are drilled and opened so
perfectly parallel that the points, when measured close
up to the brass will show the same result if moved out
a short distance, and then measured again in the same
manner.

Again, if the points or centers of the arbors are not
ground perfectly central and true with the arbor itself,
and the arbor should be turned, more or less, in moving
it in or out, the eccentricity would be brought out, and
a discrepancy occur in the measurement. Then, again,
allowing that the spacing obtained is correct, to obtain
the measurement of that spacing, a measurement must
first be made from outside to outside of these two arbors
from which the measurement of one of the arbors must
be deducted, thus leaving many chances for errors to
creep in, and create much trouble.

To get rid of these objectionable features in obtaining
the measurements of center distances, I had a pair of
parallel dividers made, which I have used with much

satisfaction and excellent results. This instrument does not differ very much from one of the same kind of Swiss manufacture, which can be obtained from almost any first class material dealer, except in some minor points, but it is made with much greater care. I am in hopes that at some period in the near future some of our enterprising watch tool manufacturers will take up this much needed branch of measuring tools for watchmakers, such for instance, as upright gauges for measuring lengths of pivots, staffs, etc., to the thousandth of an inch, and jaw gauges for the measuring of diameter, to thousandths, or even *five* and *ten* thousandths of an inch. Such gauges are now made, and are in *constant* use in all of the watch factories in America, but as yet, are not in the market on sale.

The " micrometer calipers," manufactured by Darling, Brown & Sharpe, of Providence, R. I., are splendid gauges for the purposes for which they are made, but they are not what is needed for measuring such delicate articles as watch pivots and staffs, as they are not arranged for that purpose, and are not sensitive enough. They *do* make, however, the finest instruments that were ever invented for obtaining a perfectly correct measure-

ment of center distances, diameters of wheels and pinions, etc., etc., in thousandths, or millimetres, as one
may choose to order them. Having used one of these
gauges for some years, with such satisfactory results, I
cannot speak too highly of them, or recommend them
too strongly.

For the benefit of those who may never have seen
them, I will here give their own description of them.   It
is called

"THE IMPROVED VERNIER CALIPER."

DESCRIPTION OF THE VERNIER AND ITS USE.

On the bar of the instrument is a line of inches numbered 0, 1, 2, etc., each inch being divided into ten
parts, and each tenth into four parts, making forty
divisions to the inch.   On the sliding jaw is a line
of division (called a Vernier, from the inventor's name)
of twenty-five parts, numbered 0, 5, 10, 15, 20, 25.
The twenty-five parts on the Vernier correspond, in
extreme length, with twenty-four parts, or twenty-four
fortieths of the bar, consequently each division on the
Vernier is smaller than each division on the bar by one-

thousandth part of an inch. If the sliding jaw of the caliper is pushed up to the other, so that the line marked O on the Vernier corresponds with that marked O on the bar, then the two next lines to the right will differ from each other by one-thousandth of an inch, and so the difference will continue to increase, one-thousandth of an inch for each division, till they again correspond at the line marked 25 on the Vernier. To read the distance the caliper may be open, commence by noticing how many inches, tenths and parts of tenths, the zero point on the Vernier has been moved from the zero point on the bar. Now count upon the Vernier the number of divisions, until one is found which coincides with one on the bar, which will be the number of thousandths to be added to the distance read off on the bar. The best way of expressing the value of the division on the bar,

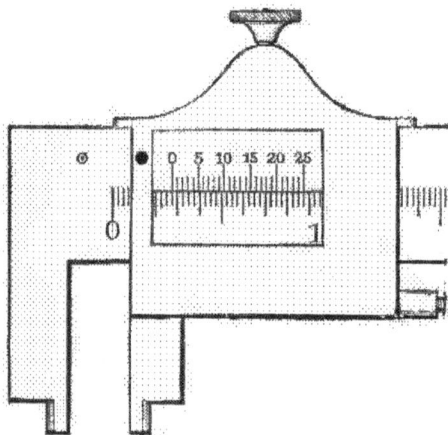

is to call the tenths one hundred thousandths (.100), and the fourths of tenths, or fortieths, twenty-five thousandths (.025). Referring to the cut above, it will be seen that the jaw is open two-tenths and three-quarters, which is equal to two hundred and seventy-five thousandths (.275). Now suppose the Vernier was moved to the right so that the tenth division should coincide with the next one on the scale, which will make ten thousandths (.010) more to be added to two hundred and seventy-five thousandths (.275), making the jaws to be open two hundred and eighty-five thousandths (.285).

In making inside measurements with the 6″ Vernier and the pocket Vernier calipers, two and one-half tenths or two hundred and fifty thousandths (.250) of an inch, and with the 12″ and 24″ Verniers, three-tenths or three hundred thousandths (.300) of an inch should be added to the apparent reading on the Vernier side for the space occupied by the caliper points. When the other side of the instrument is used, no deduction is necessary, as there are two lines, one indicating inside and the other outside measurements.

Before closing this article it may be well to give a little description of the

PARALLEL DIVIDERS.

These should be, and usually are, constructed in such a manner that the centers are interchangeable, ranging from the finest point, suitable to enter the smallest jewel hole, up to a wide, beveled center, large enough to fill the largest arbor hole in the watch. They should also be made so that the center underneath the handle, and whose shank runs up into it, is arranged as a "pump center," with a binding screw (thumb) so that it may be fastened at whatever elevation you may desire. The slide, or carriage, which runs upon the parallel bar, should be made with a "boss" on the under side, also with a binding screw, into which the other center may be fastened, this center and slide being made to be drawn forward and back, on the bar, by means of a fine-threaded screw contained within the bar.

Everything being in readiness for taking the measurements, place the center which is underneath or pumped up into the handle in one of the holes of the jewel or hole in the plate, and adjust the point or center in the slide, turning the screw either backward or forward, until that point is also located in the other jewel between which two holes you desire to obtain the distance. Then raise or lower the center in the handle until the bar is perfectly parallel with the plate. After doing this, examine your points again to see that they enter the holes perfectly or drop into them without any crowding, either to one side or the other. If this is done correctly we have the distance between the holes perfectly spaced, so that we can measure them on the Vernier caliper as often as we like, and by means of the Vernier read off the exact distance in thousandths, as often as we like, without making any change. With these two instruments there is hardly a *chance* of making any mistake.

Should you not have them, the depthing tool can be used, and the measurements obtained in the old fashioned way, but great care should be used in getting the true measurement of the desired space.

———

The following article entitled, " What is a Watch ? " was written by the writer of this article and published by the E. Howard Watch & Clock Co., in circular form, some years ago and distributed quite freely to their customers. Many commendatory letters were received in return mentioning the fact that they had found it of much value in laying before their customers the necessity of having their watches properly cared for in due season, and not to let them run so long as to make it necessary for the repairer to run up large bills in order to put them in condition to perform at all satisfactory after cleaning.

WHAT IS A WATCH?

Among the many who own and carry watches, how few ever stop to think of the amount of brain power that has been expended upon its construction, the number and delicacy of its parts, and the difficulties attendant upon the assembling of all these delicate parts into one harmonious whole, so that when completed it shall run continuously for a period of months at least, and always indicate the correct time to the very second without even a moment's rest. It is expected to be so perfectly regular in its habits, no matter how irregular we may be in ours, that it can be depended on for all our most important engagements, and should it happen to make a misstep, or stumble in the least, during its continuous run, it

will oftentimes call down upon its poor, defenceless head
the strongest anathemas.

An ordinary eighteen-size watch of the present day is
composed of about 200 pieces. Making 18,000 beats or
vibrations per hour, it has to make 432,000 per day, or
157,680,000 per year. The balance travels 1.43 inches
with each vibration, which is equal to 9.75 miles in
twenty-four hours, 292.50 in thirty days, or 3,558.75
miles in one year. The amount of oil used in oiling the
entire watch is about one-tenth of one drop, while the
oil that can be put in the balance jewels with safety
must not exceed one one-hundredth of a drop.

Should the vibrations of the balance be so accelerated,
or retarded—accelerated by means of running, jumping,
or horseback riding, etc.; or retarded by the changes
which are constantly taking place in the oil, by wear,
etc., so as to accelerate, or retard, these vibrations three
one-millionths of each vibration, the watch may gain or
lose one second per day, thirty seconds per month, or
six minutes per year.

If by any means, the little globules which form the
lubricating substance become so changed, as changed they
must be with this constant travel over them, from what they

were when the watch was first cleaned and put together
and regulated, so as to allow a little more friction on any
of the several bearings, the same result may follow.

It therefore becomes a very important matter to *know*
that the oil you use is the *very best* that can be obtained,
and put on judiciously, being *sure* that there is enough,
but *equally* sure that not so much is used as to overflow
the jewels and cause it to be drawn out upon the plates
and away from the pivots or bearings.

www.ingramcontent.com/pod-product-compliance
Lightning Source LLC
Chambersburg PA
CBHW022014190326
41519CB00010B/1520